KINDERGARTEN

NUMBERS
WORKBOOK

BEAUTY IN BOOKS

This Book Belongs to:

□————————————————————□

TRACING NUMBERS

Count the item and trace the number word.

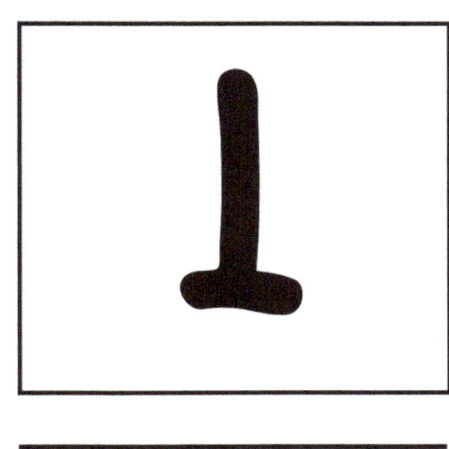

one

Trace the number.

TRACING NUMBERS

Count the items and trace the number word.

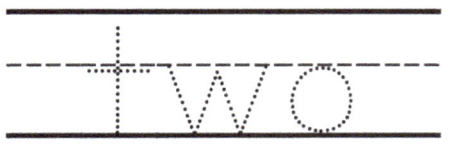

Trace the number.

2 2 2 2 2

2 2 2 2 2

2 2 2 2 2

2 2 2 2 2

TRACING NUMBERS

Count the items and trace the number word.

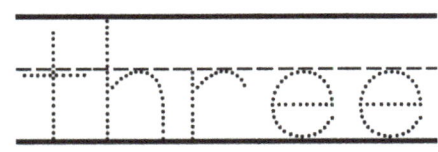

Trace the number.

3 — 3 — 3 — 3 — 3

3 — 3 — 3 — 3 — 3

3 — 3 — 3 — 3 — 3

3 — 3 — 3 — 3 — 3

Name _____   Date _____

TRACING NUMBERS

Count the items and trace the number word.

Trace the number.

TRACING NUMBERS

Count the items and trace the number word.

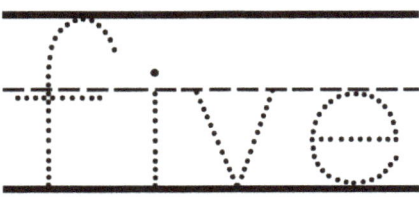

Trace the number.

5 5 5 5 5

5 5 5 5 5

5 5 5 5 5

5 5 5 5 5

TRACING NUMBERS

Count the items and trace the number word.

six

Trace the number.

6   6   6   6   6

6   6   6   6   6

6   6   6   6   6

6   6   6   6   6

Name _____   Date _____

TRACING NUMBERS

Count the items and trace the number word.

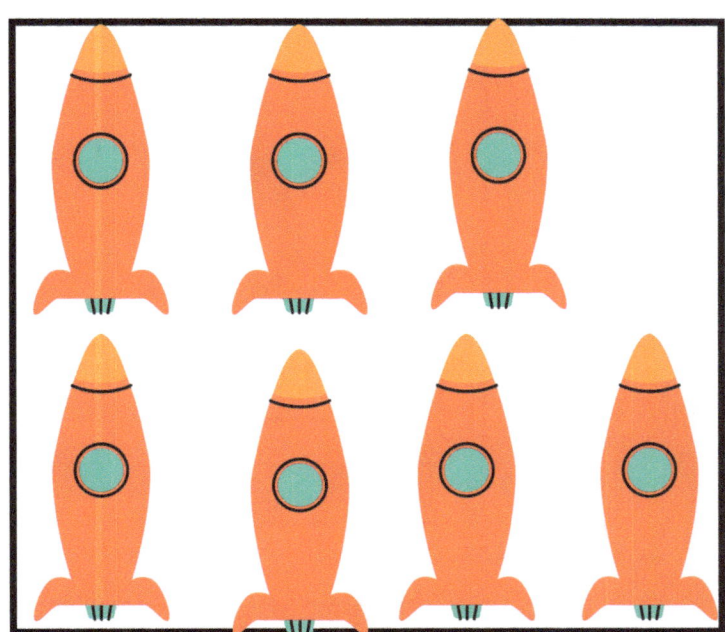

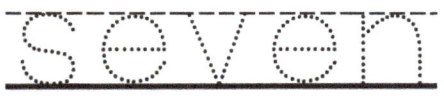

Trace the number.

7 7 7 7 7 7 7

7 7 7 7 7 7 7

7 7 7 7 7 7 7

7 7 7 7 7 7 7

TRACING NUMBERS

Count the items and trace the number word.

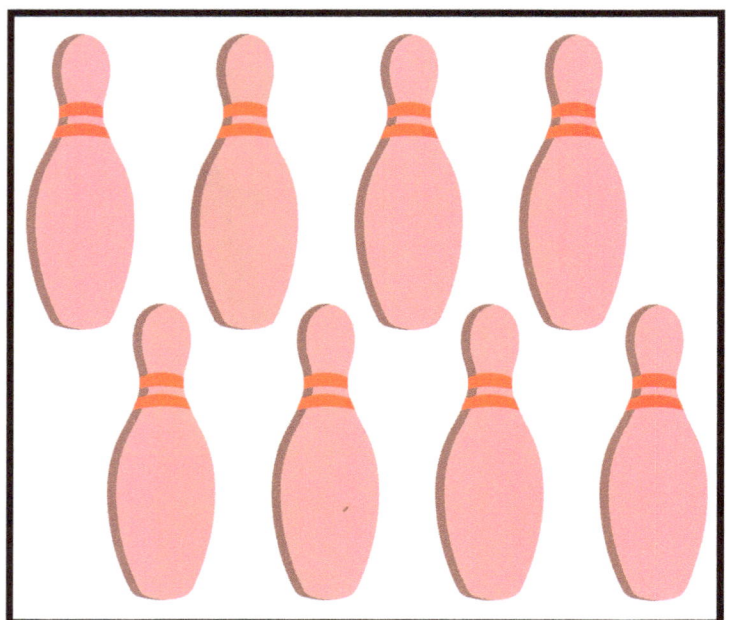

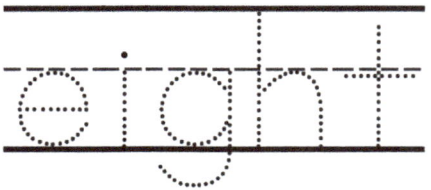

Trace the number.

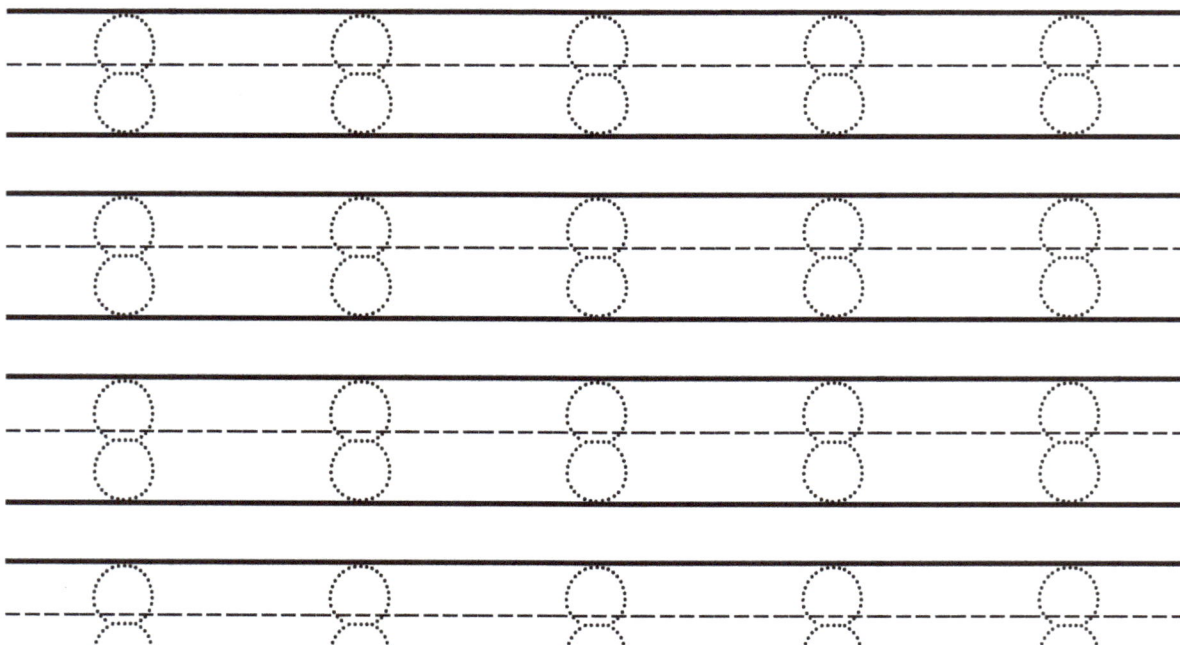

TRACING NUMBERS

Count the items and trace the number word.

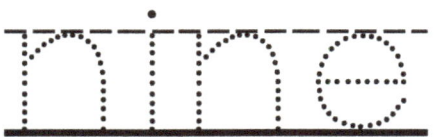

Trace the number.

TRACING NUMBERS

Count the items and trace the number word.

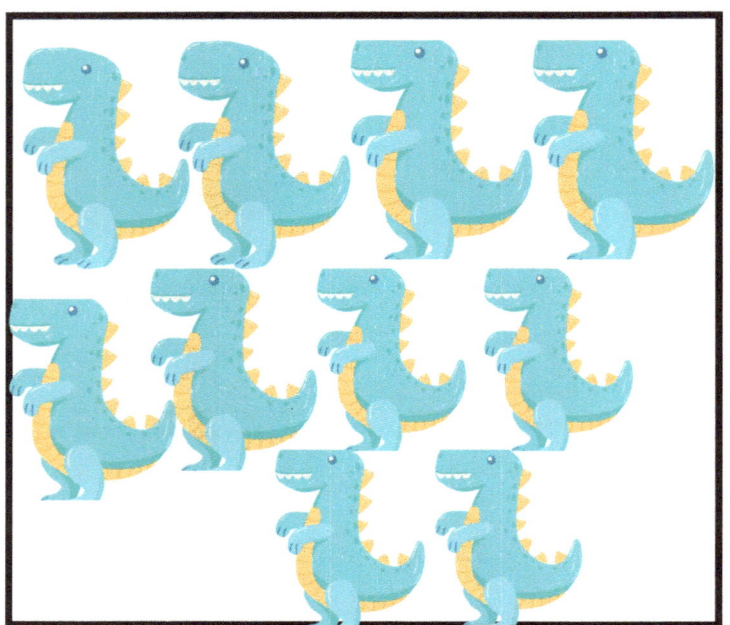

Trace the number.

Name _____   Date _____

TRACING NUMBERS

Count the items and trace the number word.

eleven

Trace the number.

TRACING NUMBERS

Count the items and trace the number word.

t̶w̶e̶l̶v̶e̶

Trace the number.

TRACING NUMBERS

Count the items and trace the number word.

Thirteen

Trace the number.

13   13   13   13

13   13   13   13

13   13   13   13

13   13   13   13

Name _____   Date _____

TRACING NUMBERS

Count the items and trace the number word.

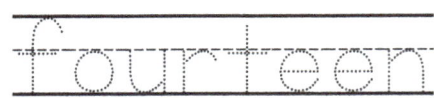

Trace the number.

TRACING NUMBERS

Count the items and trace the number word.

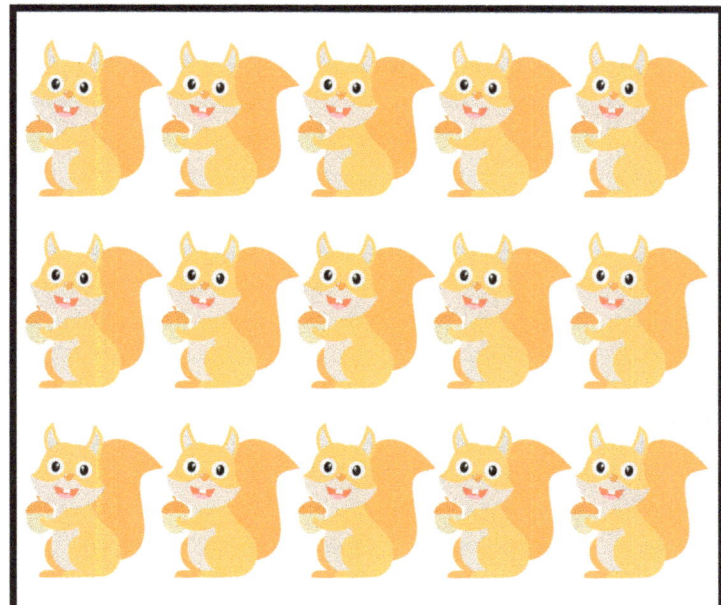

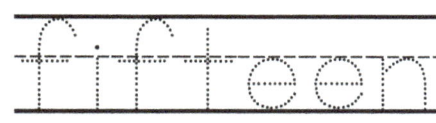

Trace the number.

Name _____   Date _____

TRACING NUMBERS

Count the items and trace the number word.

sixteen

Trace the number.

16   16   16   16

16   16   16   16

16   16   16   16

16   16   16   16

Name _____ Date _____

TRACING NUMBERS

Count the items and trace the number word.

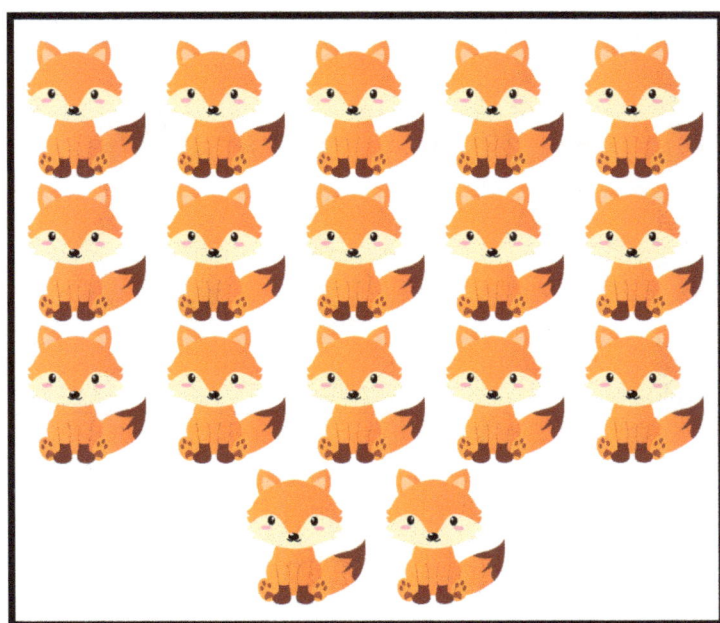

seventeen

Trace the number.

7	7	7	7
7	7	7	7
7	7	7	7
7	7	7	7

Name _____   Date _____

TRACING NUMBERS

Count the items and trace the number word.

18

eighteen

Trace the number.

18   18   18   18

18   18   18   18

18   18   18   18

18   18   18   18

Name _____   Date _____

TRACING NUMBERS

Count the items and trace the number word.

19

nineteen

Trace the number.

9   9   9   9

9   9   9   9

9   9   9   9

9   9   9   9

TRACING NUMBERS

Count the items and trace the number word.

20

Trace the number.

20 --- 20 --- 20 --- 20

20 --- 20 --- 20 --- 20

20 --- 20 --- 20 --- 20

20 --- 20 --- 20 --- 20

WRITING NUMBERS

Write the numbers 1 to 10.

1	2	3	4	5
6	7	8	9	10

Name _____   Date _____

WRITING NUMBERS

Write the numbers 11 to 20.

11	12	13	14	15
16	17	18	19	20

Name _____ Date _____

WRITING NUMBER WORDS

Trace and write the number names 1 to 10.

1	one	**one**
2	two	
3	three	
4	four	
5	five	
6	six	
7	seven	
8	eight	
9	nine	
10	ten	

Name _____   Date _____

WRITING NUMBER WORDS

Trace and write the number names 11 to 20.

11	eleven	**eleven**
12	twelve	
13	thirteen	
14	fourteen	
15	fifteen	
16	sixteen	
17	seventeen	
18	eighteen	
19	nineteen	
20	twenty	

NUMBER ORDER

Fill in the missing number in the sequence.

Name _____ Date _____

NUMBER ORDER

Fill in the missing number in the sequence.

Name _____   Date _____

NUMBER ORDER

Fill in the missing number in the sequence.

ORDERING NUMBERS

Cut and glue in order.

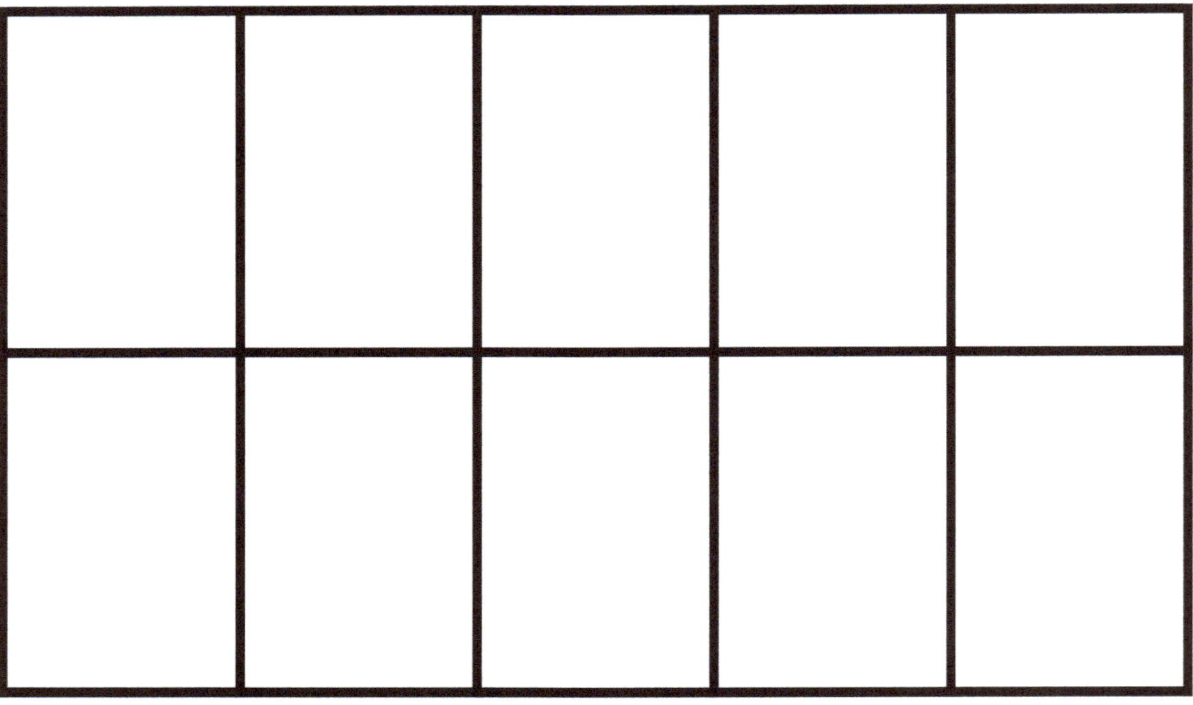

ORDERING NUMBERS

Cut and glue in order.

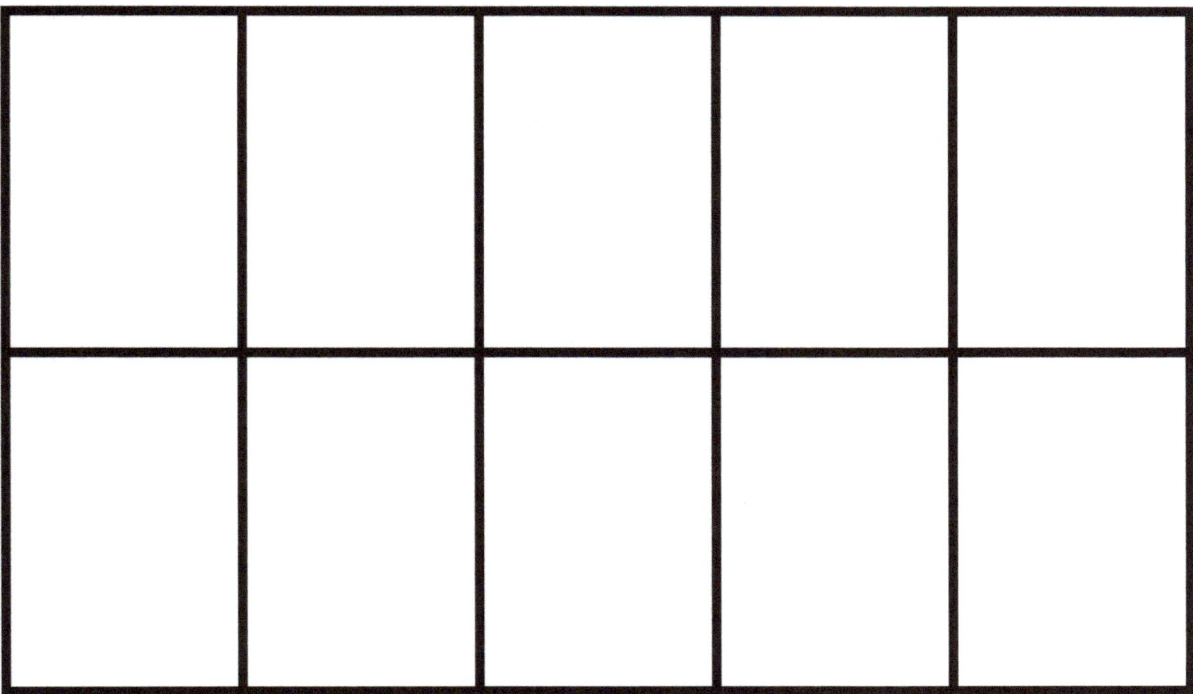

Name _____   Date _____

ORDERING NUMBERS

Cut and glue in order.

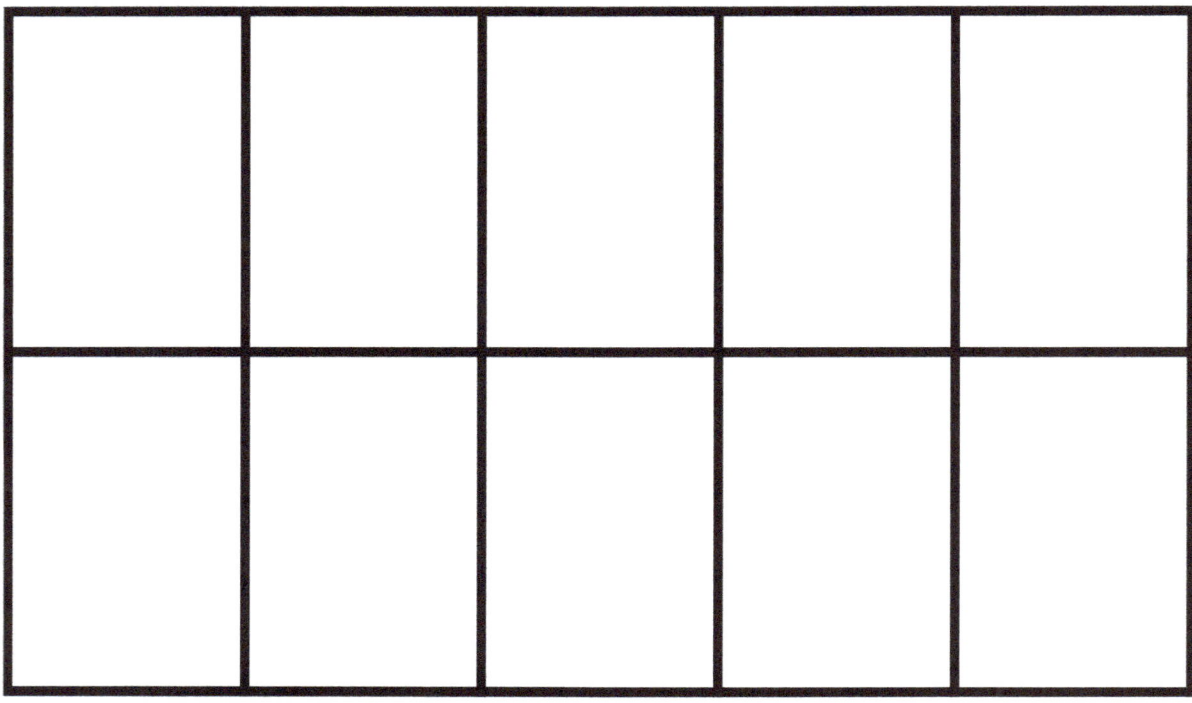

Name _____   Date _____

ORDERING NUMBERS

Cut and glue in order.

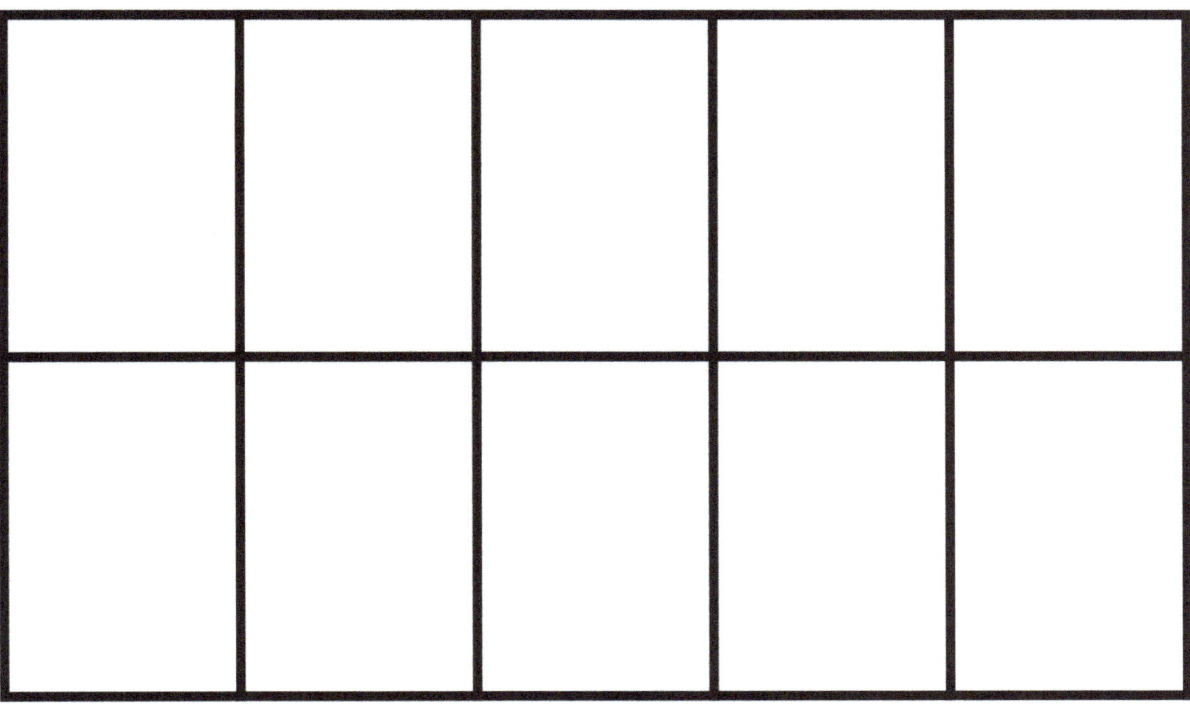

ORDERING NUMBERS

Cut and glue in order.

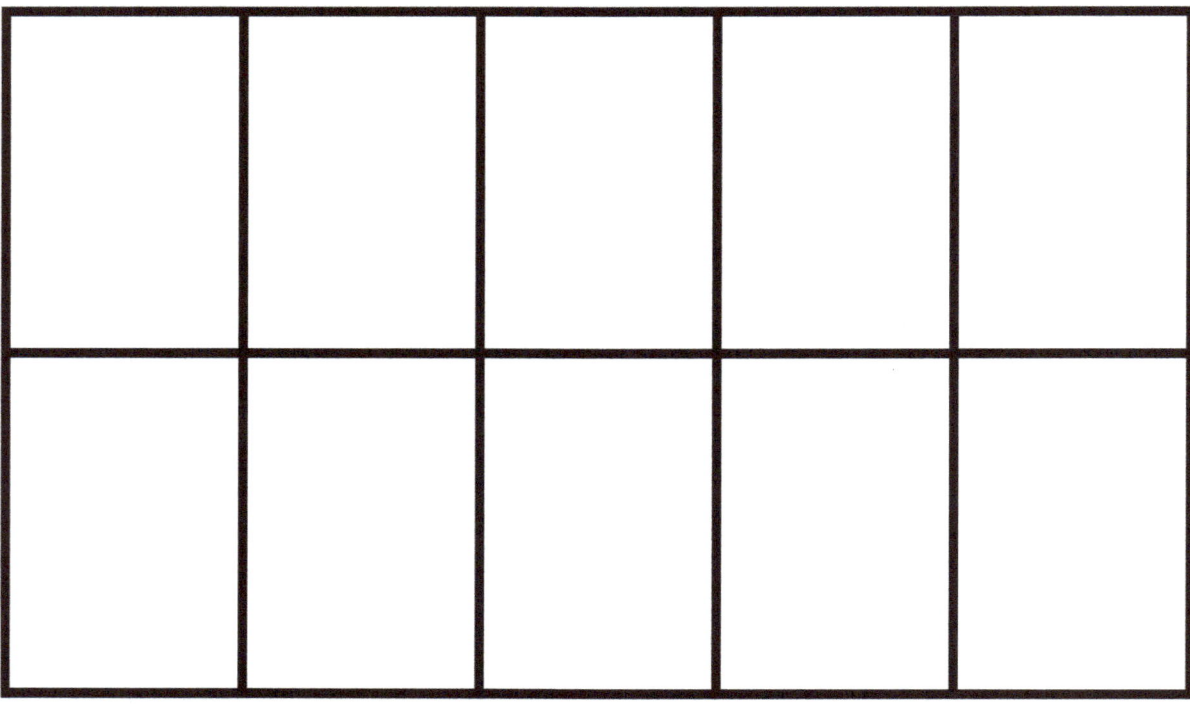

Name _____ Date _____

ORDERING NUMBERS

Cut and glue in order.

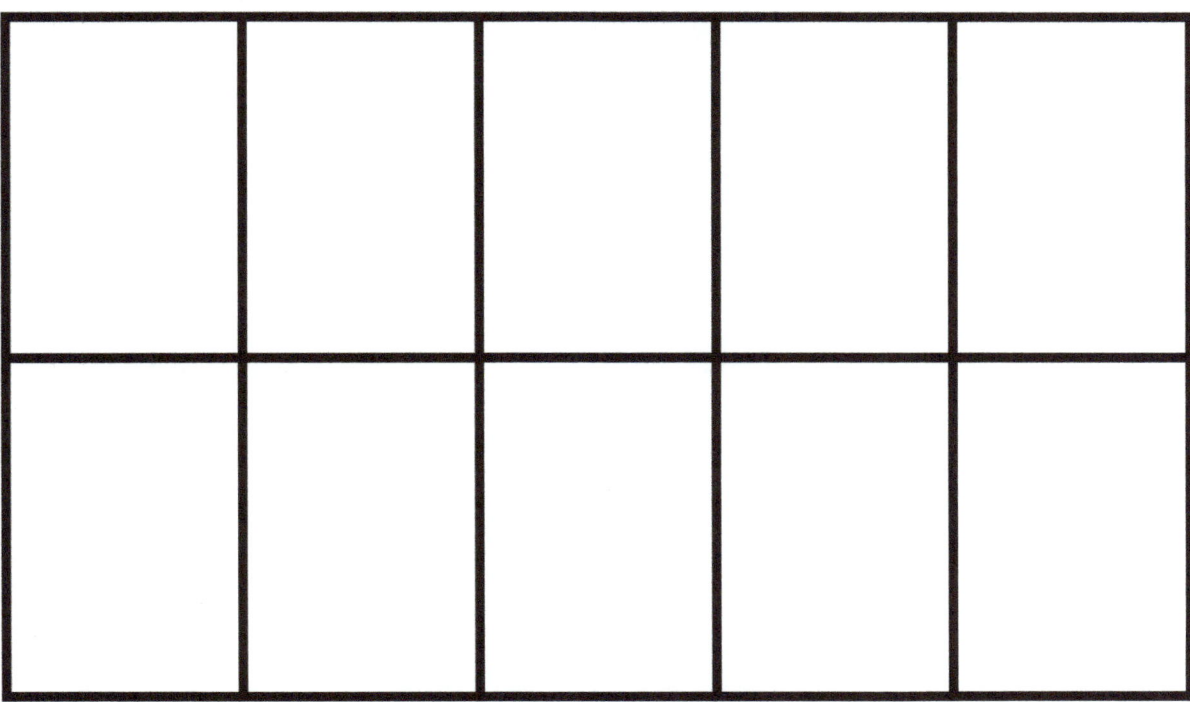

Name _____   Date _____

NUMBER BEFORE

Fill in the number that comes before.

NUMBER BEFORE

Fill in the number that comes before.

NUMBER BEFORE

Fill in the number that comes before.

NUMBER AFTER

Fill in the number that comes after.

Name _____   Date _____

NUMBER AFTER

Fill in the number that comes after.

Name _____   Date _____

NUMBER AFTER

Fill in the number that comes after.

14 ◯   12 ◯

11 ◯   15 ◯

18 ◯   13 ◯

19 ◯   10 ◯

17 ◯   16 ◯

NUMBER BETWEEN

Fill in the number that comes between.

2		4
7		9
3		5
6		8

NUMBER BETWEEN

Fill in the number that comes between.

1		3

5		7

0		2

4		6

Name _____ Date _____

NUMBER BETWEEN

Fill in the number that comes between.

8		10

9		11

13		15

10		12

NUMBER BETWEEN

Fill in the number that comes between.

11		13

15		17

12		14

14		16

12:00

Write the time on the clock.

1:00

Write the time on the clock.

2:00

Write the time on the clock.

3:00

Write the time on the clock.

4:00

Write the time on the clock.

6:00

Write the time on the clock.

7:00

Write the time on the clock.

8:00

Write the time on the clock.

9:00

Write the time on the clock.

10:00

Write the time on the clock.

11:00

Write the time on the clock.

ONE

Use a dot marker to dot circles
of the number.

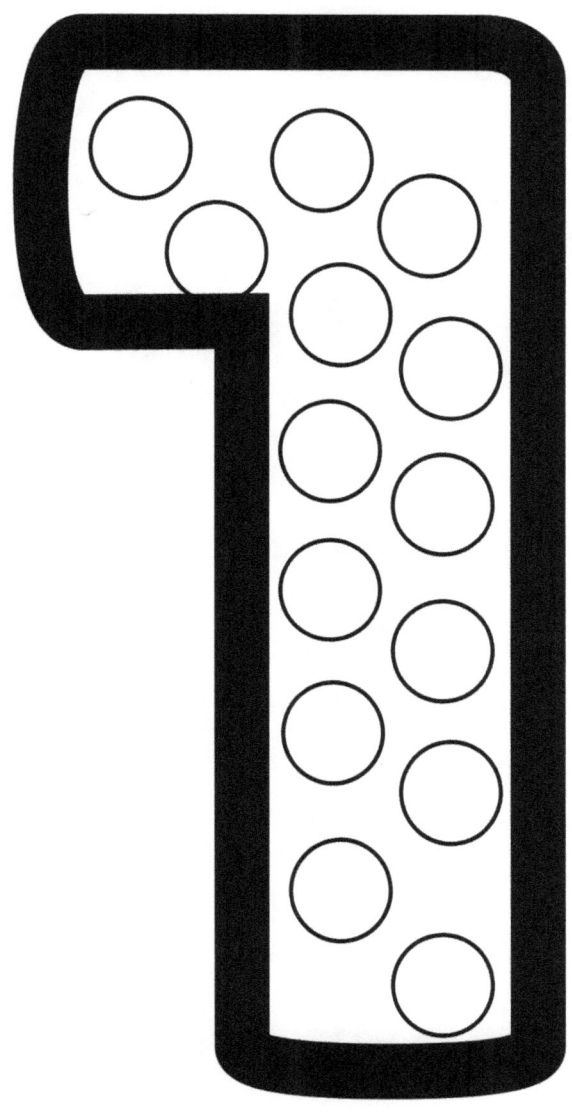

two

Use a dot marker to dot circles
of the number.

THREE

Use a dot marker to dot circles
of the number.

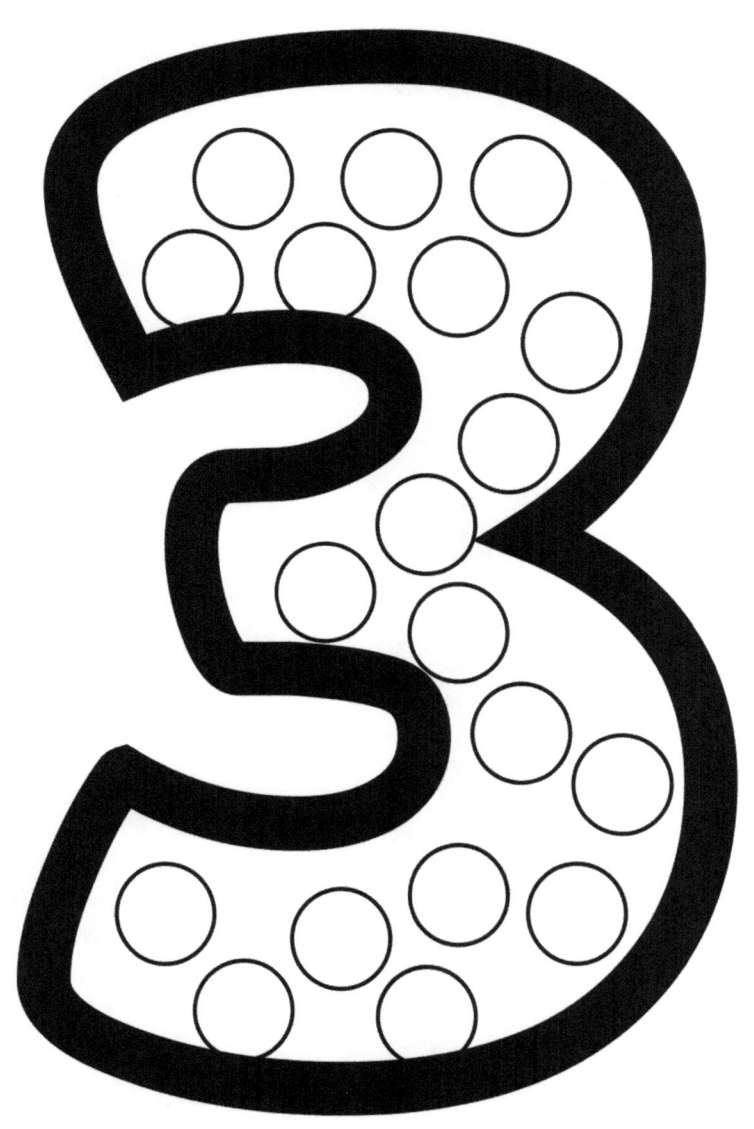

FOUR

Use a dot marker to dot circles
of the number.

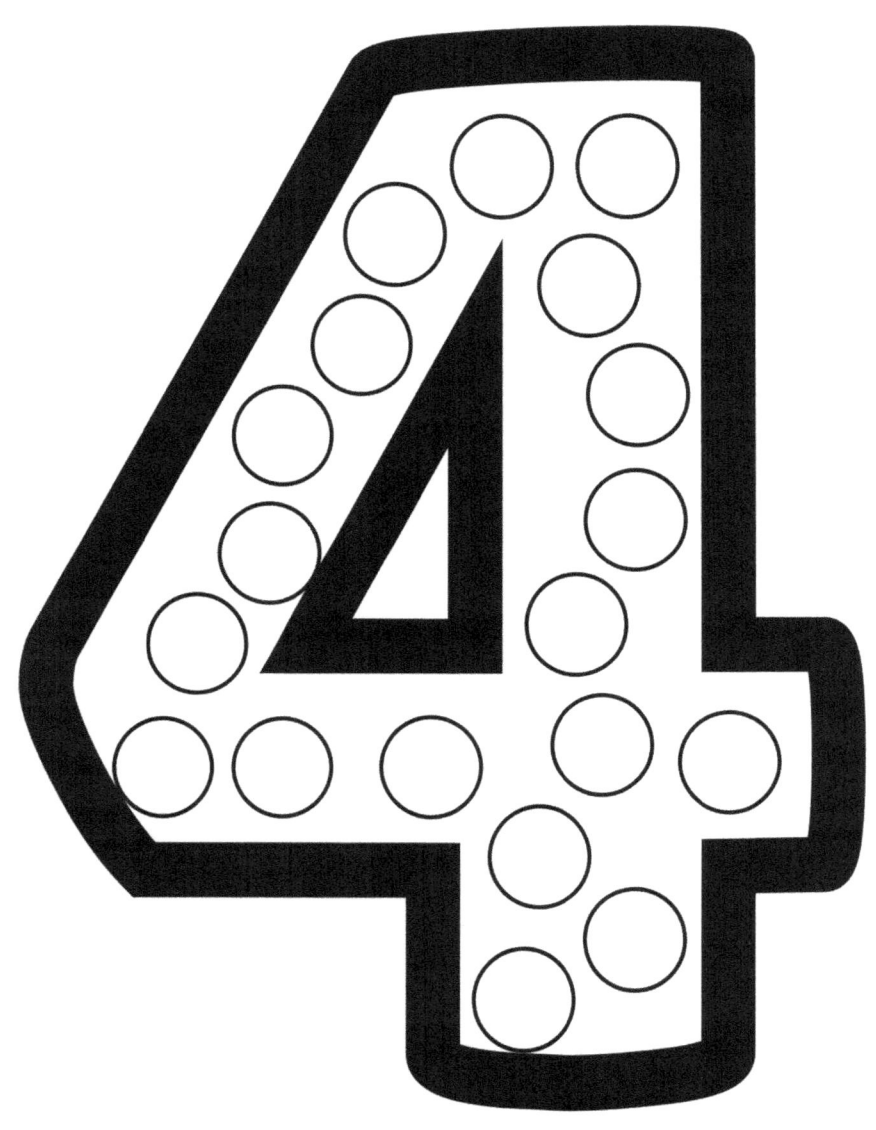

FIVE

Use a dot marker to dot circles
of the number.

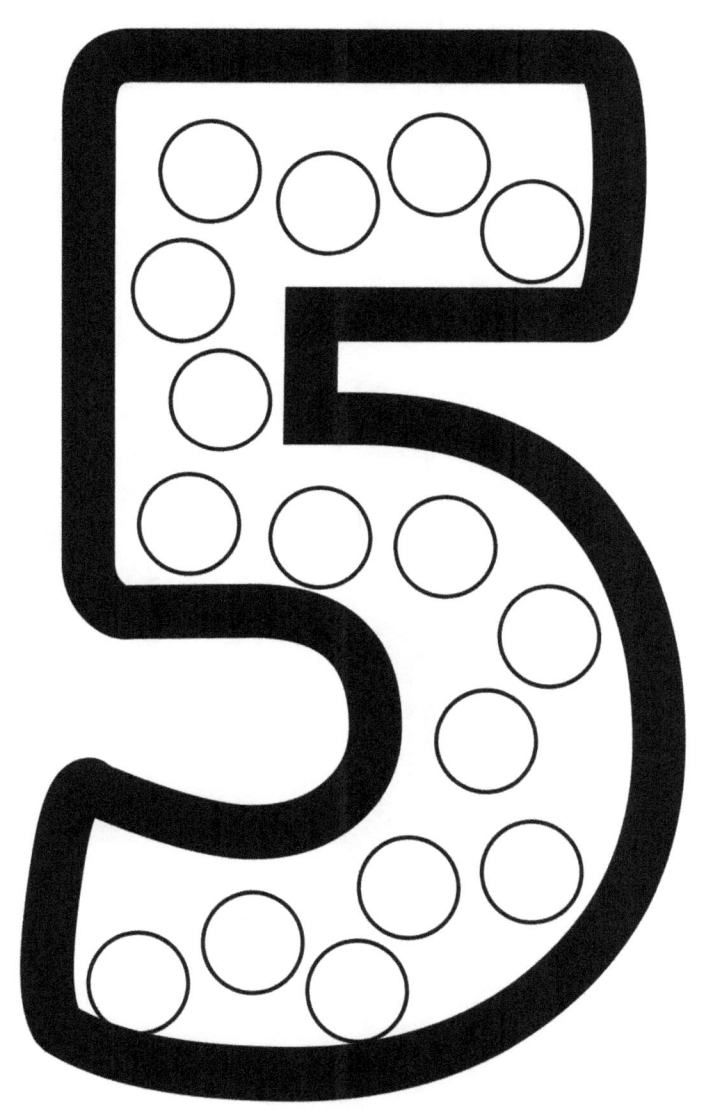

SIX

Use a dot marker to dot circles
of the number.

SEVEN

Use a dot marker to dot circles
of the number.

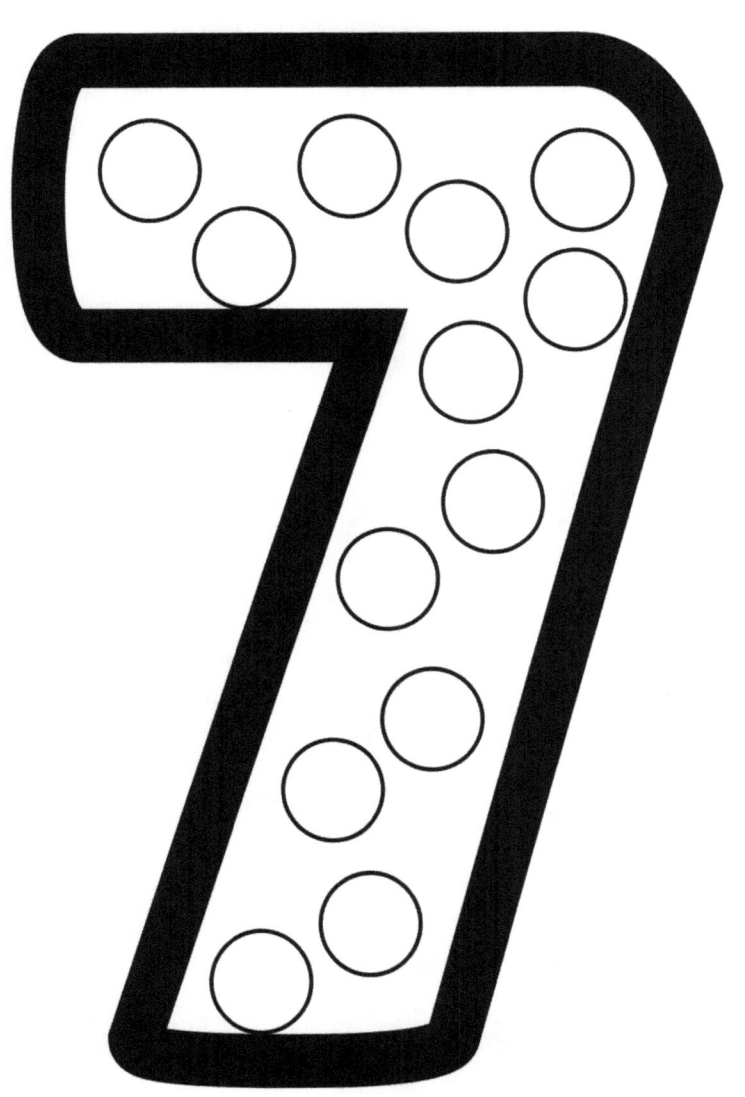

EIGHT

Use a dot marker to dot circles
of the number.

NINE

Use a dot marker to dot circles of the number.

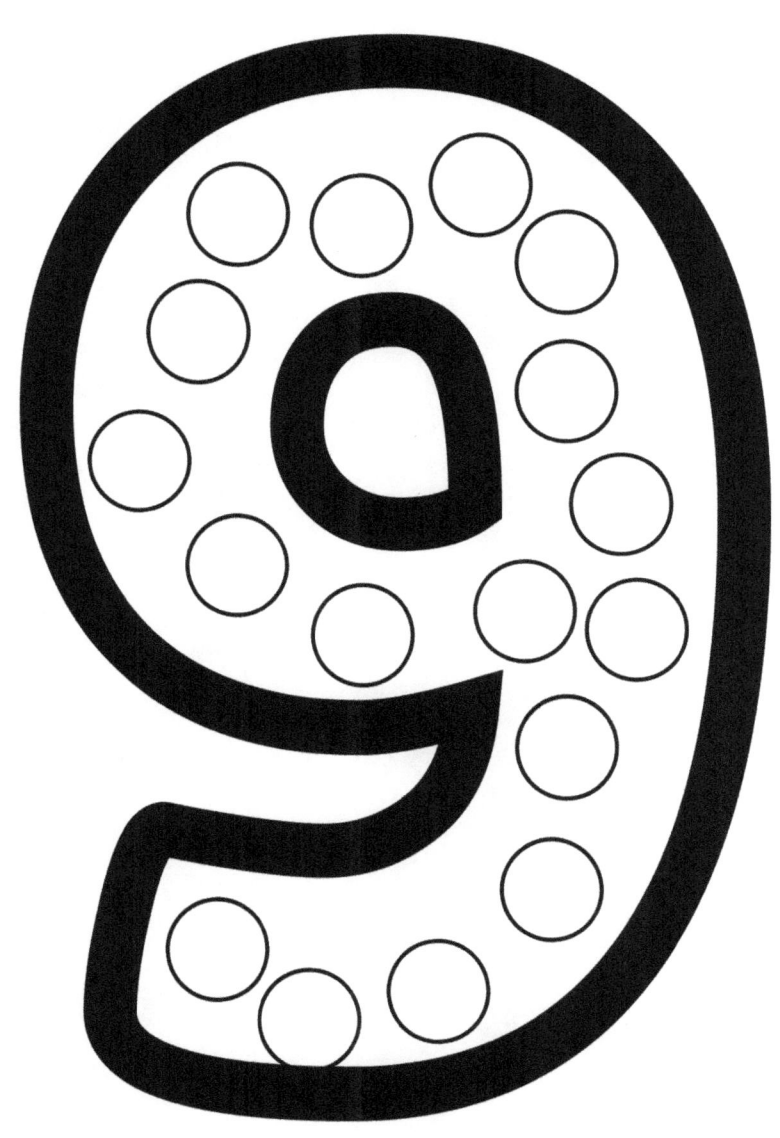

TEN

Use a dot marker to dot circles
of the number.

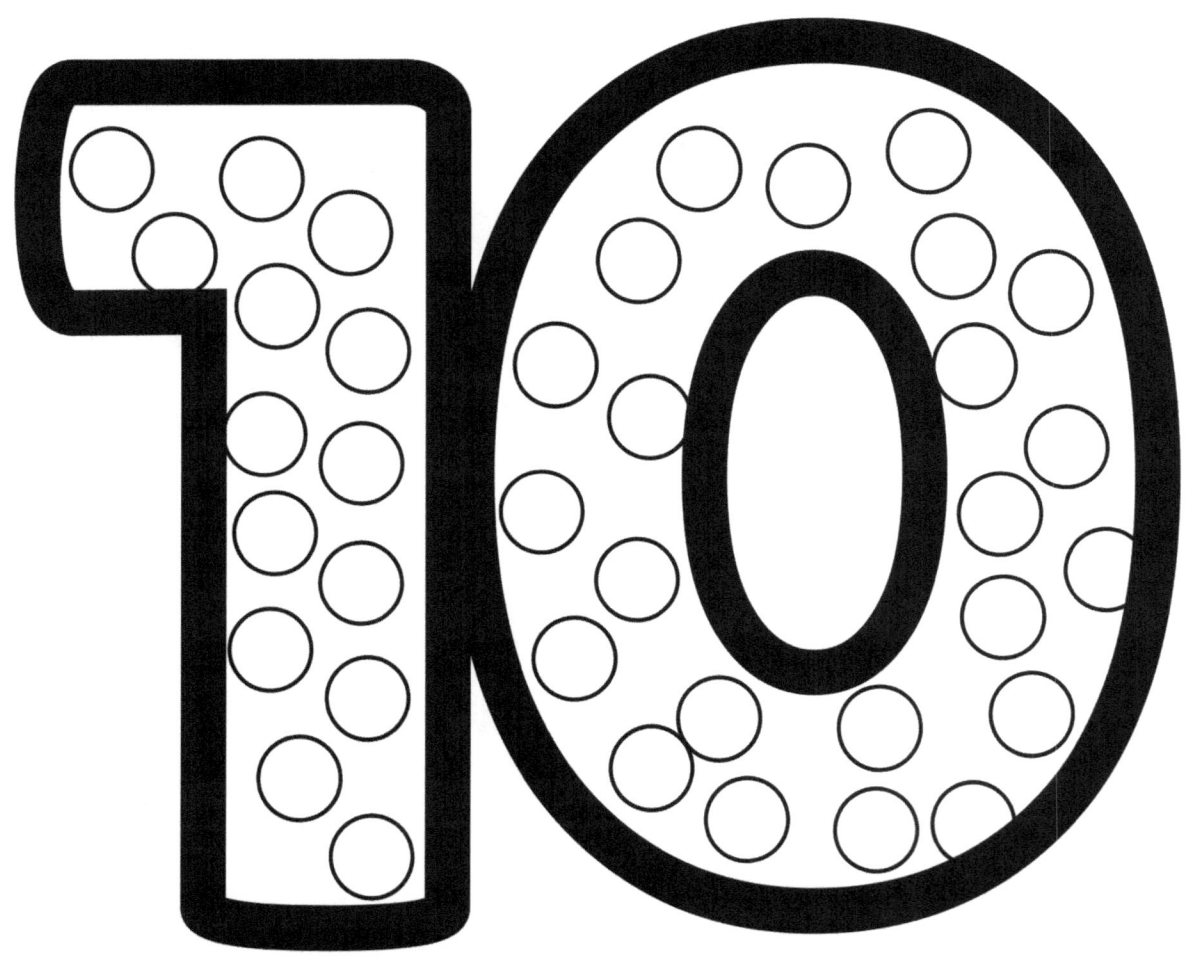

www.ingramcontent.com/pod-product-compliance
Lightning Source LLC
Chambersburg PA
CBHW040511150626
46551CB00030B/2501